MECHANICAL D[ ]

PROJECTION DRAWING.

ISOMETRIC AND OBLIQUE DRAWING.

WORKING DRAWINGS.

BY

WALTER K. PALMER, M. E.,

(OHIO STATE UNIVERSITY)

DEPARTMENT OF DRAWING, MILLER MANUAL LABOR SCHOOL,

CROZET, VIRGINIA.

British Library Cataloguing-in-Publication Data
A catalogue record for this book is available from the
British Library

Technical Drawing and Drafting

Technical drawing, also known as 'drafting' or 'draughting', is the act and discipline of composing plans that visually communicate how something functions or is to be constructed.

It is essential for communicating ideas in industry, architecture and engineering. The need for precise communication in the preparation of a functional document distinguishes technical drawing from the expressive drawing of the visual arts. Whereas artistic drawings are subjectively interpreted, with multiply determined meanings, technical drawings generally have only one intended meaning. To make the drawings easier to understand, practitioners use familiar symbols, perspectives, units of measurement, notation systems, visual styles, and page layout. Together, such conventions constitute a visual language, and help to ensure that the drawing is unambiguous and relatively easy to understand.

There are many methods of constructing a technical drawing, and most simple among them is a sketch. A sketch is a quickly executed, freehand drawing that is not intended as a finished work. In general, sketching is a quick way to record an idea for later use, and architects sketches in particular (in a very similar manner to fine artists) serve as a way to try out different ideas and establish a composition before undertaking more finished work. Architects drawings can also be used to convince clients of the merits of a design, to enable a building constructer to use them, and as a record of completed work. In a similar manner to engineering (and all other technical drawings), there is a set of conventions (i.e particular views, measurements, scales, and cross-referencing) that are utilised.

As opposed to free-sketching, technical drawings usually utilise various manuals and instruments. The basic drafting procedure is to place a piece of paper (or other material) on a smooth surface with right-angle corners and straight sides – typically a drawing board. A sliding straightedge known as a 'T-square' is then placed on one of the sides, allowing it to be slid across the side of the table, and over the surface of the paper. Parallel lines can be drawn simply by moving the T-square and running a pencil along the edge, as well as holding devices such as set squares or triangles. Other tools can be used to draw curves and circles, and primary among these are the compasses, used for drawing simple arcs and circles. Drafting templates are also utilised in cases where the drafter has to create recurring objects in a drawing – a massive time-saving development.

This basic drafting system requires an accurate table and constant attention to the positioning of the tools. A common error is to allow the triangles to push the top of the T-square down slightly, thereby throwing off all the angles. Even tasks as simple as drawing two angled lines meeting at a point require a number of moves of the T-square and triangles, and in general drafting this can be a time consuming process. In addition to the mastery of the mechanics of drawing lines, arcs, circles (and text) onto a piece of paper – the drafting effort requires a thorough understanding of geometry, trigonometry and spatial comprehension. In all cases, it demands precision and accuracy, and attention to detail.

Conventionally, drawings were made in ink on paper or a similar material, and any copies required had to be laboriously made by hand. The twentieth century saw a shift to drawing on tracing paper, so that mechanical copies could be run off efficiently. This was a substantial development in the drafting process – only eclipsed in the twenty-first century with 'computer-aided-drawing' systems (CAD). Although classical draftsmen and women are still in high demand, the mechanics of the drafting task have largely been automated and accelerated through the use of such systems. The development of the computer had a major impact on the methods used to design and create technical drawings, making manual drawing almost obsolete, and opening up new possibilities of form using organic shapes and complex geometry.

Today, there are two types of computer-aided design systems used for the production of technical drawings; two dimensions ('2D') and three dimensions ('3D'). 2D CAD systems such as AutoCAD or MicroStation have largely replaced the paper drawing discipline. Lines, circles, arcs and curves are all created within the software. It is down to the technical drawing skill of the user to produce the drawing – though this method does allow for the making of numerous revisions, and modifications of original designs. 3D CAD systems such as Autodesk Inventor or SolidWorks first produce the geometry of the part, and the technical drawing comes from user defined views of the part. This means there is little scope for error once the parameters have been set.

Buildings, Aircraft, ships and cars are now all modelled, assembled and checked in 3D before technical drawings are released for manufacture.

Technical drawing is a skill that is essential for so many industries and endeavours, allowing complex ideas and designs to become reality. It is hoped that the current reader enjoys this book on the subject.

CONTENTS.

ISOMETRIC AND OBLIQUE.

WORKING DRAWINGS.

PREFACE.

The following pages contain the substance of a progressive course, beginning with the essential principles of elementary Projection Drawing, and passing on as rapidly as is consistent with thoroughness, through Isometric and Oblique Drawing, to the making of *working drawings*.

The work is simply outlined, briefly, for the convenience of teacher and students. Explanations and illustrations by the teacher will be needed, and the course may be supplemented or modified by him, as the needs of a particular class may seem to require.

The little book is meant to be simply a "teacher's help,"—a text strictly. It is, however, not a text, such as the usual text books on Mechanical Drawing. It embodies a general method of instruction. The aim is to assist the student in developing for himself the essential fundamental principles, in a natural and progressive order, starting with the most elementary ideas, and working along by easy steps from one fact to another, until the whole subject is unfolded.

The principles are in most cases simply stated. The student should thoroughly verify each in order, with the assistance of the teacher when necessary. Explanations and illustrations are not given, for each statement follows directly and easily from what has just preceded it, and it is one of the chief benefits of the work, to verify the statements given and make simple sketches and models to illustrate them.

By little helps in the way of hints and illustrations, and by questioning to recall facts already established, any student may be led to develop for himself all the essential principles of the subjects treated, in such a way that he cannot well forget them. At the same time he acquires the habit of working from a knowledge of principles,—of thinking and reasoning, and relying on what he knows.

The plates given are chosen to illustrate the principles brought out. Coming as they do at the close of each topic, after the principles have been established, they bring those principles into use, and fix the method of their application, thus preparing for the next step.

No drawings are shown, and as few figures as possible are used, as it is expected that the teacher will supply what is needed to clear up individual difficulties, and that he will develop the actual work of drawing, in the class room, by little helps and suggestions.

It is left altogether with the teacher as to what shall be drawn under the head of "Working Drawings." It is well to draw from measurements taken directly from objects, or from sketches taken from machine details. Shop exercises in a technical school afford good practice. Revising a set of badly made shop drawings, is good training. The work should be arranged as progressively as possible.

Considerable practice should be given in making the general drawings of complete machines, and detailing from them; also in detailing from drawings made by others; after which a course in designing may properly be given.

A scheme for tabulating and indexing the details, should be arranged by the student when he makes a general drawing of a machine, and this should be followed by those who detail from his drawing.

It will be necessary for the teacher to supply much relating to working drawings, in the way of conventional methods of doing various things, such, for instance, as representing screw threads, drawing hexagonal bolt heads, etc. Such things are best given to students individually, as they are ready to make use of them. There are many little "ways" that must be brought out by the teacher, as the need for them occurs. He will of course draw upon standard works of reference, his own experience, and reliable examples of practice. Students should be frequently referred to books by good authorities, and to shop drawings from the drafting rooms of leading machine works, manufacturing companies and railroad shops.

Great pains should be taken to have thoroughly good drawings made. *Accuracy, neatness* and *correctness* should be insisted upon. With proper attention to details, really fine work can be obtained from average students.

A standard size of plate is recommended, until the working drawings are reached. A small size of plate is preferable, as working on a small drawing is found to enforce accuracy. "Freedom" can be quickly acquired when the student comes to a large drawing. If too much "freedom" be allowed at first, accuracy will never be acquired.

Only the very best paper should be used. A standard plate of Keuffel & Esser's "normal" paper, $6\frac{3}{4}$ x $8\frac{1}{2}$ ", with a border line $\frac{3}{4}$ " from the edge all around, has proved very satisfactory.

Working drawings should be penciled upon ordinary "detail paper," then traced upon good tracing cloth, and blue prints made from the tracings, if time allows.

When the student is done, the exercises and plates,

and the working drawings he has made, which should be thoroughly good work, and can easily be, with proper attention on the part of the teacher, make with the text a complete work upon the subject. Thus the student helps to make for himself a work of reference for his permanent use, and while so doing becomes thoroughly grounded in the principles and practices of the work.

The method is not only natural and systematic, but trial in the class room proves that it arouses and maintains the interest of the student, and eventually produces the most satisfactory results in the way of knowledge and skill acquired, and training in systematic work and study.

It is assumed that the student is reasonably skillful in the use of instruments before taking up this course, and it should be preceded by a thoroughly good course in Geometric Construction Drawing. A previous introduction into Geometry is necessary, as a knowledge of the terms and definitions of Geometry is assumed.

W. K. P.

Crozet, Va., April, 1894.

MECHANICAL DRAWING.

INTRODUCTION.

1. **Mechanical Drawing** is the art of making drawings capable of representing mechanical and architectural structures, and the parts composing them, so clearly and completely that skilled mechanics can make those structures exactly as they are intended to be, without any further directions than those contained in the drawings themselves.

2. Such drawings made expressly for the workmen are called "**Working Drawings.**"

3. Evidently, these drawings, to meet the requirements, must express easily and perfectly all facts in regard to the *position, form* and *magnitude* of objects represented.

4. In other words, they must be capable of expressing the *geometry* of all mathematical figures,—solid as well as plane,—for **Geometry** is defined as "that branch of mathematics which treats of *position, form* and *magnitude.*"

5. They must represent these solid figures *in space*, yet the drawings must all be in one plane—that of the paper.

6. It is evident, that the *art* of Mechanical Drawing must have as its foundation an exact mathematical science. This is the science of "**Projections.**"

7.   Drawings made in accordance with the methods of Projection Drawing, meet all requirements completely.

REVIEW QUESTIONS.

Q. 1.   What is meant by Mechanical Drawing?
Q. 2.   What are "Working drawings?"
Q. 3.   What are the requirements to be met by working drawings?
Q. 4.   What is the definition of Geometry?
Q. 5.   What is the distinction between an art and a science?
Q. 6.   Is Mechanical Drawing a science or an art?   Or is it both?
Q. 7.   What is the foundation of Mechanical Drawing?
Q. 8.   How are the requirements of Mechanical Drawing met?

PROJECTION DRAWING.

PRINCIPLES OF PROJECTION.

8. **Projection Drawing** is the science and art of producing drawings which shall represent completely all facts of position, form and magnitude of all geometrical quantities in space.

9. The **Methods** employed in Projection Drawing, are those of "**Orthographic Projection**," which is the basis of the science of *Descriptive Geometry*.

FUNDAMENTAL IDEAS OF PROJECTION.

10. In Projection Drawing, all objects are represented by their *projections* upon fixed planes of reference. Hence an understanding of the geometrical meaning of the term "*projection*" is essential.

11. From Geometry we have : " The **projection of a point** upon a straight line is the *foot* of the perpendicular dropped from the point upon the line."
Thus :

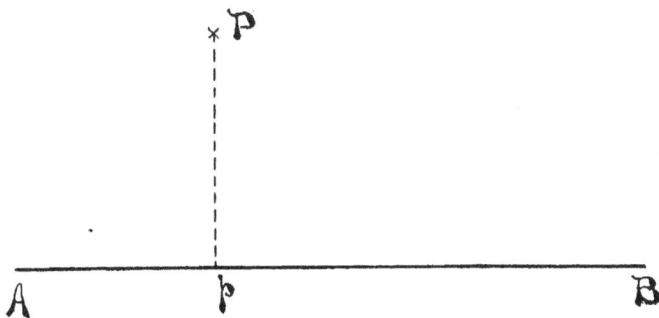

Figure 1.

Let AB be any straight line, and P any point not in the line. Let Pp be perpendicular to AB. Then p is the *projection* of P upon the line AB.

12. Similarly :   '' The **projection of a point** upon a **plane** is the foot of a perpendicular from the point to the plane. Thus :

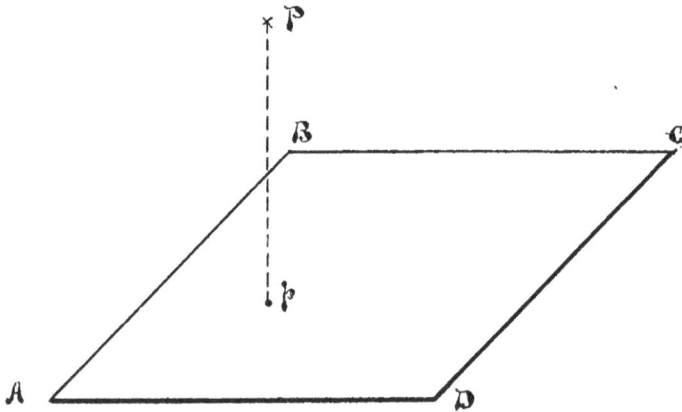

Figure 2.

P is a point in space, and Pp a perpendicular from P upon the plane $ABCD$. p is the *projection*, then, of P upon the plane.

13. **The projection of any line**, or **surface**, or **solid**, may be found by finding the projections of all of its points. For lines and surfaces are but successions of points, and solids are bounded by surfaces.

APPLICATION OF PROJECTION TO DRAWING.

14. Now consider two mathematical planes of indefinite extent, intersecting each other at right angles, fixed in position, one horizontal and the other vertical. Let the following figure represent limited portions of these two planes :

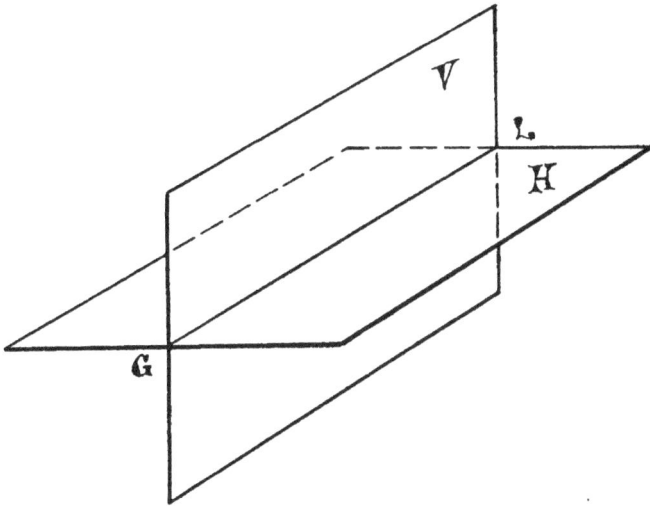

Figure 3.

Let the vertical plane be denoted by V and the horizontal plane by H, and their line of intersection by GL.

15.   Now imagine a point to be situated in space in one of the angles of these two planes, as shown in Fig. 4 :

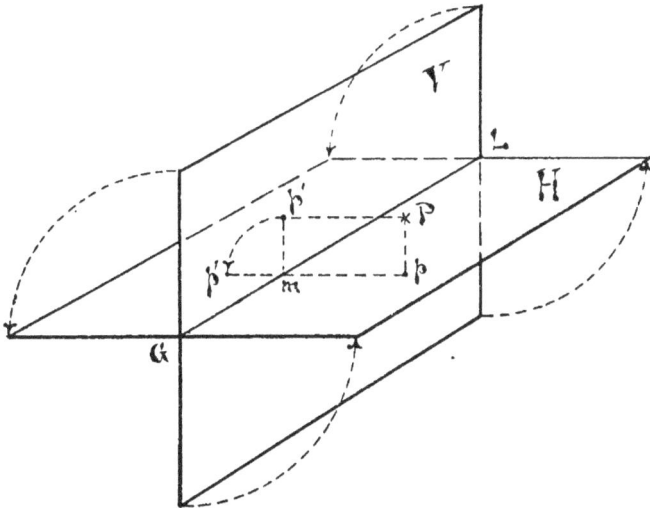

Figure 4.

Let P be the point in space in the angle of the two planes.

Now *project* P upon the horizontal plane, and then on the vertical plane. Then p is the projection of P upon the horizontal plane, and p' is the projection of P upon the vertical plane.

Draw pm and $p'm$ perpendicular to GL, in the horizontal and vertical planes respectively.

16. Now it can readily be seen that these *projections* of P upon the two planes, fix the position of P in space, exactly. For if p and p' be known, P can be found at once by erecting perpendiculars, which will intersect in the position of P.

17. By means, then, of their projections upon two *fixed planes*, points anywhere in space may be represented. If *points* can be thus represented, lines and surfaces, and all objects, may be represented in the same way.

18. Now, while by this means all geometrical quantities can be completely represented in space, no application can be made of this method until some means is provided for bringing everything into one plane, which can be made to coincide with the surface of a sheet of drawing paper.

19. From Fig. 4 it is easily seen that all the quantities may be readily brought into one plane, and without destroying any of the essential relations.

Imagine the vertical plane to revolve backward about its intersection, GL, while the horizontal plane remains fixed. Let it revolve downward until it coincides with the horizontal plane, carrying its projection of the point with it to a new position p'.

20. Evidently all quantities will retain their same relations, and we may now deal with the revolved positions of the point, and no longer consider the point.

21. When the two planes have been made to coincide, they, with the projections of the point, may be brought into the plane of the paper, and will appear as shown in Fig. 5:

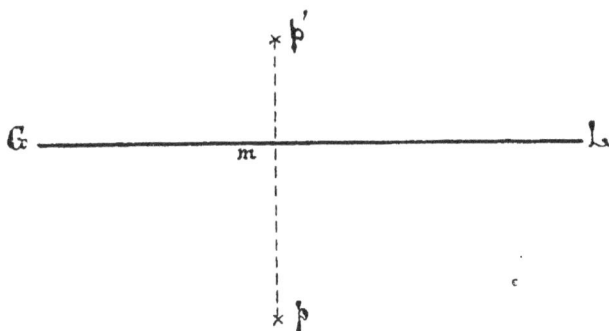

Figure 5.

The line of intersection GL, is placed horizontal on the drawing, with the projections upon the vertical plane above, and those upon the horizontal plane below it.

NOTATION.

TERMS EMPLOYED.

22. For convenience and brevity the following terms are used:

(1) The two fixed planes are called "*the planes of projection*," or "*the co-ordinate planes*," or "*the planes of reference*."

(2) The vertical plane of projection is called simply "*the vertical plane*," or commonly, "*V*."

(3) The other is called "*the horizontal plane*," or simply, "*H*."

(4) The line of intersection of the two planes is called "*the ground line*," or briefly, "*GL*."

(5) The projection upon V (p', Figs. 4 and 5), is called "*the vertical projection*," or simply "*the V projection.*"

(6) The projection upon H (p, Figs. 4 and 5), is called "*the horizontal projection*," or "*the H projection.*"

(7) The perpendiculars dropped from a point in space (Pp and Pp', Fig. 4), are known as *projecting lines*. On the actual drawing, Fig. 5, pm and $p'm$ are commonly called the projecting lines, though they are really the *projections* of the projecting lines.

LETTERING.

23. (1) Points in space are denoted by a capital letter, as P, Fig. 4.

(2) Projections of points are lettered with the corresponding small letters.

(3) Vertical projections are always *primed*.

(4) Horizontal projections are not.

(5) If a point in space be moved into several positions in succession, the same letter is used for all of the positions. The successive positions of the V projection are denoted by the same letter that is used for the first, marked with ^{II}, ^{III}, ^{IV}, and so on in order. The H projections are marked with the corresponding subscripts, $_2$, $_3$, $_4$, &c.

For example, if a point A should move successively to A^{II}, A^{III}, A^{IV}, &c., its V projections would be lettered a^{I}, a^{II}, a^{III}, a^{IV}, &c., and its H projections a, a_2, a_3, a_4, &c.

(6) Projections of plane figures and all surfaces and solids, are lettered by selecting prominent points on them, and lettering these according to the system of lettering points in general.

(7) Projections of particular angles are marked by small dotted arcs, and lettered with the minor letters of the Greek alphabet. But one letter is used to denote an angle. The ^{I}, ^{II}, ^{III}, &c., and subscripts, are used on letters marking the projections of angles, just as on letters denoting projections of points.

GENERAL PRINCIPLES.

POINTS.

24. By reference to Fig. 4, the following fundamental principles are established in regard to the projections of points, on the actual drawing, where the two planes coincide, as in Fig. 5:

(1) The two projections of a point in space fall on a straight line, which is perpendicular to the ground line. The vertical projection is *above*, the horizontal *below* GL.

(2) The distance of the *vertical* projection from the ground line, is equal to the distance of the point from the horizontal plane in space.

(3) The distance of the *horizontal* projection from the ground line, is equal to the distance of the point from the *vertical* plane in space.

LINES.

25. As lines are made up of successions of points, *any* line may be represented by finding the two projections of *all* its points. These projections of the points constitute the *projections of the line*.

SURFACES AND SOLIDS.

26. In general, the two projections of surfaces and solids may also be considered as made up of the projections of *all* the points composing them.

27. In the case of limited portions of surfaces, it is sufficient to consider simply the projections of the *lines* bounding them.

28. In the case of solids, the projections of the *bounding surfaces* only are necessary.

29. Definite principles governing the projections of lines, surfaces and solids, and special directions for particular cases, will be given as occasions arise for their use.

POINT OF SIGHT.

30. Consider the two projections of an object. It is evident that either projection is just what would be seen by an observer looking at the object in a direction perpendicular to the plane of projection, if the eye were at an infinite distance from the plane, and endowed with correspondingly infinite power.

31. **The point of sight** is thus assumed at an infinite distance so that the lines of sight will coincide with the projecting lines of the object.

32. The vertical projection may then be considered as a side or front view of the object. And in the same way, the horizontal projection may be considered the "top view."

33. When the point of sight is assumed in this way at an infinite distance, the method of representation is known as *Orthographic Projection*.

34. When the point is at any position within finite distance, the method is known as Scenographic, and the resulting projection is called the *Perspective* of the object.

35. Perspective views are *perfect pictures*, but are not suitable for working drawings, as the dimensions of objects are not given by them in their true relation.

DRAWING.

CONVENTIONAL LINES.

Ordinary Full Line.

36. The projections of lines in space, either given or required, are drawn with an ordinary "weight" of "*full line.*"

Auxiliary Line—Fine Full Line.

37. Projections of lines that are auxiliary,—that is used simply as *aids*,—are drawn with full lines made enough finer than the ordinary "full lines" to be in marked contrast with them.

Construction Line.

38. The two projections of all points are always joined with a "*dotted line*," made up of short dashes,—best about $\frac{1}{16}$ of an inch long and $\frac{1}{32}$ of an inch apart,—of the weight of the "fine full line."

This is also used to indicate the paths of moving points, as in cases of revolution, where the projections of points move from one position to another.

Invisible Line.

39. Projections of lines which are hidden from view by intervening surfaces, are drawn with the "*invisible line*," which is made up of dashes about $\frac{3}{32}$ of an inch long, and $\frac{1}{32}$ of an inch apart, and of the weight of the "fine full line."

Ground Line.

40. The ground line is drawn with a heavy full line, about twice the weight of the ordinary full line.

ISOLATED POINTS.

41. Projections of isolated points are marked with small fine crosses, as in Fig. 5.

EXERCISE.

42. Prepare a plate of specimens of the conventional lines to be used. Adopt lines of the proper "weight" and kind, and mark each line with its name. Leave space upon the plate for two or three additional lines, which will be met with later.

Keep this plate for reference, and always make all lines according to it.

PROJECTIONS OF A POINT.

43. The two projections of any point in space may be readily constructed directly from the fundamental principles in regard to projections of points—page 13.

PLATE I.

Represent by their two projections the six points situated as follows :

(1) A point $\frac{1}{2}$ " from V and $1\frac{3}{4}$ " from H.

(2) A point in V, and 1 " from H.

(3) A point 1 " from V, and $\frac{1}{4}$ " from H.

(4) A point $\frac{1}{2}$ " from V, and $1\frac{1}{4}$ " from H.

(5) A point $\frac{3}{4}$ " from V, and in H.

(6) A point $\frac{1}{8}$ " from V, and $\frac{7}{8}$ " from H.

Directions : Draw the ground line thro the plate a little below the middle, stopping it at equal distances from the end border lines. Place the points so that their projecting lines on the drawing will be equal distances apart. Letter the projections correctly.

Remark : For a plate 6¾ " by 8½ " with ¾ " border, the ground line should be 2½ " from the lower border, stopping ½ " from each end border line.

PROJECTIONS OF LINES.

44. In general the projections of any line, curved or straight, may be found by projecting all of its points,

but in reality *all* the points can never be projected.

45. In the case of a **curved line**, a number of points are chosen, as near together as the accuracy of the work may demand, and their two projections found. The corresponding projections of these two points are then joined by smooth curves, which are the projections of the curve in space.

46. In the case of a straight line, it is sufficient to simply project any two of its points, and then join the corresponding projections of these points by straight lines.

PROJECTIONS OF RIGHT LINES.

47. The following principles in regard to the projections of **right lines** are of great importance. They may be easily verified by a consideration of the lines in space, situated according to the assumptions made. They should be illustrated by simple models.

PRINCIPLES.

(1) The projections of a line can never be longer than the line itself.

(2) When a line is parallel to either co-ordinate plane, its projection on that plane is equal to the actual length of the line in space. Its projection on the other plane is parallel to the ground line.

(3) If a line is parallel to both co-ordinate planes, it is parallel to the ground line, and both of its projections are parallel to the ground line.

(4) If a line is perpendicular to either co-ordinate plane, its projection on that plane is a straight line perpendicular to the ground line.

(5) If a line is oblique to one co-ordinate plane, but is parallel to the other, the projection on the plane to which it is parallel gives the true size of the angle

which the line in space makes with the plane to which
it is oblique.

Evidently, from the preceding, *when a line is oblique to both planes,
neither projection gives its true length, or the angle it makes with either
plane.*

(6) If a point be in a line, or a line is to pass through
a given point, the projections of the point must be on
the projections of the line.

(7) If two planes in space intersect, their projections
intersect, and the two projections of the point of intersection lie on a straight line, perpendicular to the
ground line.

For the point of intersection is in both lines, hence its projections are
on the projections of both, and therefore at their intersection. The
point of intersection is simply a point, hence its projections are governed
by the laws of points in general.

(8) If two lines are parallel in space, their corresponding projections are parallel.

Principles (6) and (7) are equally true for lines in general, as well as
for *straight* lines.

PLATE II.

Construct the two projections of each of the three
straight lines situated as follows :

(1) A line, *CK*, 1¾ '' in length, parallel to *V*, and
parallel to *H*, 1⅝ '' above *H*, and 1⅛ '' in front of *V*.

(2) A line *BH*, 2'' in length, parallel to *H* and inclined to *V* at an angle of 45°, the line to be 1⅛ '' above
H, and the extremity nearest *V*, ½ '' from *V*.

(3) A line *FL*, 1⅝ '' in length, parallel to *V*, and inclined at an angle of 30° to *H*, ¾ '' in front of *V*, and its
lower extremity 1 '' above *H*.

Directions : Draw *GL* as before. Arrange the
drawing to suit the plate.

For standard 6¾ by 8½ plates, the directions are : *Place the first projecting line ⅞'' from the left border. Leave ½'' between the projecting
lines of (1) and (2), and of (2) and (3).*

REVOLUTION.

48. It often becomes necessary to revolve points, and lines, and even entire figures, in space, by means of their projections. Evidently, any object may be revolved, if the principles governing the revolution of a *point* are known.

REVOLUTION OF A POINT.

49. A point is said to be **revolved** about a straight line, as an axis, when it is so moved that it describes the arc of a circle, whose center is the axis, and whose plane is perpendicular to the axis.

50. The angle through which the point is revolved, is measured by the arc described.

51. It is evident, that when all points of a figure are revolved about an axis, and through the same angle, their *relative positions* remain unchanged.

PRINCIPLES.

52. The following are true in regard to the revolution of points. They are readily verified by considering the quantities in space :

(1) If a point be revolved about an axis, perpendicular to one plane of projection, its projection on that plane describes a circle arc. Its projection on the other plane moves in a straight line parallel to the ground line.

(2) If the axis be parallel to one plane of projection, and oblique to the other, the projection of the point of the plane to which the axis is parallel, moves in a straight line perpendicular to the projection of the axis on that plane. The other projection describes an ellipse.

It is a general rule that when the axis is parallel to one plane, however situated with respect to the other, that the projection of the point on the plane to which the axis is parallel, moves in a straight line perpendicular to the projection of the axis.

PLATE III.

I.   Given the two projections of a point P, situated $1\frac{5}{8}$ " above H, and $\frac{7}{8}$ " in front of V, and an axis perpendicular to H, 1 " from the point, revolve the point about the axis through an angle of 75°, and find its projections in the revolved position.

II.   Given the point M in space, situated $\frac{7}{8}$ " above H, and $1\frac{1}{2}$ " from V.   Revolve it, first about an axis perpendicular to V, $1\frac{1}{8}$ " from it, through an angle of 90°, and then from this position, about an axis perpendicular to H, $\frac{3}{4}$ " from it, revolve it through an angle of 60°, and construct its projections.

Directions :   Draw *GL* as before.   Arrange the drawing to suit the plate.   The axes being *given quantities*, will be drawn with full lines.   Letter all the projections correctly.   For the projections of the radius of revolution in each case, use the "dotted line " here.

REVOLUTION OF LINES.

53.   When a straight line is oblique to both planes of projection, neither projection shows its true length, or the angle it makes with either plane.

54.   To find the **true length** of a line so situated, or the **angles** it makes with the planes of the projection, it is necessary to *revolve* the line.

55.   **To revolve** a straight line, it is only necessary to revolve any two of its points through the same angle, and join their revolved positions by a straight line. Whenever possible, the axis of revolution is assumed through some point of the line,—preferably one extremity.   Then it is only necessary to revolve one point,—usually, the other end.

PLATE IV.

Given the two projections of a line, the left-hand end of which is $1\frac{1}{4}''$ above H, and $\frac{3}{8}''$ in front of V, and the right-hand end $\frac{1}{2}''$ above H, and $1\frac{7}{8}''$ in front of V, its H projection being inclined 45° to GL.

Find the true length of the line, and the angles it makes with V and with H.

Directions: Draw GL as before. Arrange the drawing in the middle of the plate.

The assumed, and derived projections, are *given* and *required*, hence are drawn with the ordinary full line. The lettering must show the order of the steps taken.

The projections of the axes of revolution, will not be drawn. They will be omitted hereafter. But they should be sketched for the study of problems.

The ground line may be left unlettered hereafter.

PLATE V.

Find the two projections of a line $1\frac{7}{8}''$ long, situated so that its left-hand end is $\frac{3}{4}''$ above H, and $1\frac{1}{4}''$ in front of V, and its right-hand end is $1\frac{3}{8}''$ above H, and $1\frac{7}{8}''$ in front of V.

Directions: Draw GL as usual. Place the problem in the middle of the plate.

Suggestion: It will be necessary to place the line first in an *auxiliary* position.

REVOLUTION OF PLANE FIGURES.

56. In the case of **polygons**, it is only necessary to revolve their vertices, according to the principles of revolution of points in general. All points of a polygon must be revolved through the same angle.

57. In the case of a **plane curve**, points must be assumed upon the curve as near together as the accuracy of the work demands, and these revolved, all thro the same angle, according to the rules for points in general.

PLATE VI.

Having given the two projections of a regular hexagon, in space, parallel to V, two sides of the hexagon parallel to H, about an axis perpendicular to H, thro the left-hand extremity of the horizontal diameter, revolve the hexagon thro an angle of 45°, and construct its projections.

Then revolve it again about the same axis, till the plane of the figure inclines 60° to V. Revolve again, till the hexagon comes perpendicular to V.

Directions: Draw GL as before. Let the diameter of the circumscribing circle of the hexagon be 2 ″.

Letter the projections of the vertices of the hexagon in all their positions.

PLATE VII.

Having given a circle 2 ″ in diameter, in space, parallel to H, about an axis perpendicular to V, tangent to the circle, revolve it downward until its plane is inclined at an angle of 45° with H. Then assume an axis thro the center of the circle, perpendicular to H, and revolve the circle about it through an angle of 30°.

Construct the projections in both positions.

Directions: Draw GL as usual. Arrange the drawing to suit the plate.

PROJECTIONS OF SOLIDS.

58. The projections of solids may now be readily constructed, from the principles of projection of lines and surfaces.

PLATE VIII.

I. Construct the two projections of a right regular hexagonal prism, altitude $2\frac{1}{2}$ ″, radius of base $\frac{3}{4}$ ″,—the base of the prism parallel to H, and two faces of the prism parallel to V.

II. Revolve the prism thro an angle of 30° to the right, keeping one point in the base at the same height above *H*, and the two faces parallel to *V*.

Construct the projections in the revolved position.

Directions: Draw *GL* whenever necessary to allow space for the constructions.

Transfer the projections of the revolved position to the right of those of the first, to avoid confusion.

SHADE LINES.

59. On drawings of solid objects, it is customary to indicate by means of *"shade lines"* the surfaces on which light falls, and on which no light falls.

CONVENTIONAL DIRECTION OF LIGHT.

60. The light is assumed to come always from over the left shoulder of the observer, in parallel straight lines, in the direction of the body diagonal of a cube, situated with two of its faces parallel to *V*, and two parallel to *H*.

61. The *projections*, then, of a line, or "ray" of light, both make angles of 45° with the ground line.

DEFINITION OF SHADE LINES.

62. The projections of all lines of an object, which separate light from dark surfaces, are *shade lines*.

63. They are drawn with heavy lines, of the weight used for the ground line.

64. Shade lines, properly placed, add greatly to the appearance of a drawing, and make it easier to understand or "read."

65. The shade lines of a drawing are, in most cases, easily determined by considering the direction of light, and how the rays must strike upon the object represented. In some instances, however, it is difficult to

decide which should be shade lines, without a knowledge of the methods of finding the shadows of objects.
It is, however, rarely difficult in ordinary mechanical
drawing, to determine the shade lines.

EXERCISE.

66. Place the proper shade lines on the projections
of Plate VIII.

PLATE IX.

Construct the two projections of the following :

(1) A right regular hexagonal pyramid, base parallel to H, two edges of the base parallel to V,—radius
of base $\frac{3}{4}$ ", altitude of the pyramid $2\frac{1}{2}$ ".

(2) A right circular cylinder, axis perpendicular to
H, diameter $1\frac{1}{4}$ ", altitude $2\frac{1}{2}$ ".

(3) A right circular cone, axis perpendicular to H,
diameter of base $1\frac{1}{4}$ ", altitude $2\frac{1}{2}$ ".

Directions: Place all three objects the same height
above H, and with their axes the same distance from
V.

Place the proper shade lines.

PLATE X.

Construct the two projections of the following :

(1) A sphere 2 " in diameter, situated in space.

(2) A prolate ellipsoid of revolution, major axis
$2\frac{1}{2}$ ", minor axis $1\frac{1}{2}$ ".

Directions: Arrange the drawing to suit the plate.
Decide in regard to *shade lines*.

Note. An *ellipsoid of revolution* is generated by revolving an ellipse
about one of its axes. If about the major axis, a *prolate* ellipsoid is generated,—if about the minor axis, an *oblate* ellipsoid.

THIRD PROJECTION.

67. In certain special cases, the usual two projections do not completely represent an object. For in-

stance, if a cylinder should lie with its axis parallel to the ground line, the *H* and *V* projections would not represent it, as they would not then be different from the two projections of a square *prism* similarly situated.

68. In every such case a *third* projection is necessary. This is obtained as follows :

69. A plane is "passed" perpendicular to both *H* and *V*, in any convenient position, just to the right of the object to be represented.

70. This plane is indicated by the lines in which it intersects *H* and *V*. These lines are called "traces," and are drawn with a line made up of a dash and three dots repeated,—the dashes about $\frac{3}{16}$" long, and of the weight of the full line,—the dots like those of the construction line,— which is the Descriptive Geometry symbol for "Auxiliary Plane Trace."

71. This plane is known as the "**profile plane.**" Having the third co-ordinate plane, the object is projected upon it, and it is revolved to the right about its vertical trace until it coincides with *V*. Then *V* is revolved down as usual, carrying the projection of the profile plane with it.

72. By means of these three projections, every possible geometric figure can be represented completely.

SHADE LINES.

73. Shade lines on the third projection are arbitrarily placed, just as they are in the vertical projection.

PLATE XI.

Construct the projections of a hollow cylinder, outside diameter $1\frac{1}{4}$ ", inside diameter $\frac{3}{4}$ ", length $2\frac{1}{2}$ ", situated with its axis parallel to *H* and to *V*, $\frac{7}{8}$ " in front of *V*, and $1\frac{1}{2}$ " above *H*. ·

Directions: Draw the ground line wherever necessary. Arrange the drawing to suit the plate.

For the standard 6¾ " by 8½ " plate, *GL* should be 2¼ " above the lower border.

REVIEW QUESTIONS.

Q. 8.   What is Projection Drawing?

Q. 9.   Its methods are those of what science?

Q. 10   Of what great branch of mathematics is Orthographic Projection the basis?

Q. 11.   In Projection Drawing, how are objects represented?

Q. 12.   What is the "projection" of a point upon a line?

Q. 13.   Upon a plane? Illustrate.

Q. 14.   How is the projection of a line found?

Q. 15.   Of a surface? Of a solid?

Q. 16.   How may two projections of a point be made to fix the position of the point in space?

Q. 17.   How are the two fixed planes assumed to be situated in space?

Q. 18.   If points are fixed by their two projections, may lines be also?

Q. 19.   And surfaces, and solid figures?

Q. 20.   Can any use be made of this method of representation, with the planes actually at right angles?

Q. 21.   What is necessary in order that application may be made of it?

Q. 22.   How is this requirement met?

Q. 23.   Will the two projections fix a point as well when the two planes are made to coincide?

Q. 24.   How is the plane figure resulting from bringing the two planes into coincidence, placed on the drawing? Make a sketch showing the two projections of a point.

Q. 25.   What are the two fixed planes called?

Q. 26.   What is the name of the vertical plane?

Q. 27.   Of the horizontal?

Q. 28.   Of the line of their intersection?

Q. 29.   What is meant by the *H* projection of a point?

Q. 30.   The *V* projection?

Q. 31.   What are projecting lines?

Q. 32.   How are the projections of a point lettered?

Q. 33.   How are the projections of the successive positions, *P, P''*, *P''', P*ᴵⱽ, &c, of a point, lettered?

Q. 34.   How are the projections of plane figures lettered?

Q. 35.   Of solids?

Q. 36.   How are projections of important angles marked?

Q. 37.   How are the two projections of a point always situated with respect to the ground line?

Q. 38.   How is the distance of a point from *H* shown?

Q. 39.   Which projection shows the distance of a point in space from *V*?

Q. 40. How may the projections be considered as " views ? "

Q. 41. Which projection is the top view ?

Q. 42. What is the difference between orthographic and scenographic projection ?

Q. 43 What is the scenographic projection of an object commonly called ?

Q. 44. What are conventional lines ?

Q. 45. What are the ones used in Projection Drawing ?

Q. 46. How are the projections of curved lines found ?

Q. 47. Of straight lines?

Q. 48. Can the project of a straight line be longer than the line itself ?

Q. 49. When will the projection equal the line ?

Q. 50. If a line is parallel to one plane and oblique to the other, what is known regarding the projections ?

Q. 51. Which projection gives the angle with the plane to which it is oblique ?

Q. 52. If a line is parallel to H and V, how are its projections situated ?

Q. 53. If a line is perpendicular to one plane, what is its projection on that plane ?

Q. 54. What is its other projection, and how situated ?

Q. 55. If a point be on a line, straight or curved, how are the projections of the point and line situated with respect to each other ?

Q. 56. If two lines, straight or curved, intersect, what is known of their projections ?

Q. 57. If two straight lines are parallel in space, what is known of their projections ?

Q. 58. When is a point said to be revolved about a straight line as an axis ?

Q. 59. How must all points of a figure be revolved, not to change the figure ?

Q. 60. If a point moves about an axis perpendicular to H, how does its V projection move ?

Q. 61. What is the general statement of the principle ?

Q. 62. How does the projection of the point on the plane to which the axis is parallel move, in all cases, whether the axis is perpendicular to the other plane or not ?

Q. 63. Of what use are the principles of revolution ?

Q. 64. How is a straight line revolved ?

Q. 65. How is the true length of a line, oblique to both planes, found ?

Q. 66. If a line is inclined to both H and V, how is its angle with H found ?

Q. 67. How would a polygon in space be revolved ?

Q. 68. Any curve ?

Q. 69. For what are shade lines used ?

Q. 70. What is the conventional direction of light ?

Q. 71. Are the rays of light assumed to be parallel ? Why ?

Q. 72. How are the projections of a ray of light inclined to the ground line ?

Q. 73. What lines are shade lines ?

Q. 74. How are they drawn ?

Q. 75. Of what use are shade lines ?

Q. 76. How are shade lines determined on a drawing ?

Q. 77. Do two projections of all objects always represent them complete ?

Q. 78. What is a notable instance ?

Q. 79. What is done when the usual two projections are not sufficient ?

Q. 80. How is the third plane passed ?

Q. 81. When the projection has been formed on it, what is done with it ?

Q. 82. How is this third plane shown on the drawing ?

Q. 83. What is the kind of line used for the " traces ?"

Q. 84. What is the third plane called ?

Q. 85. Can everything be completely represented by three projections ?

Q. 86. How are shade lines placed on the third projection ?

ISOMETRIC AND OBLIQUE DRAWING.

DEFINITION.

74. **Isometric Drawing** is a method of representing solid objects so that their three principal dimensions will be shown in their true values by means of but one view.

75. An isometric drawing of an object is an approximate *picture* of it, which, unlike a true perspective view, shows the length, breadth and thickness in their correct values.

ISOMETRIC PROJECTION.

76. Consider a cube in space, in the angle of the two planes of projection, with its base parallel to the horizontal plane, and one diagonal of the base parallel to the vertical plane.

77. Now let the cube be revolved forward from the vertical plane, about an axis parallel to the ground line coinciding with the diagonal of the base, until a body diagonal of the cube comes perpendicular to the vertical plane.

78. It is readily seen now, that in this position the vertical projection of the edges of the cube, which are, of course, all equal in space, will be equal. For, all the edges make equal angles with the diagonal, or its direction, which is perpendicular to V, and hence equal angles with V. Hence V projections are equal.

79. This V projection is the **Isometric projection** of the cube.

ISOMETRIC DRAWING.

80.　In the isometric projection of a cube, the projections of the edges are all less, of course, than the edges themselves in space; but all the edges are equally "foreshortened,"—that is, the projections are *proportional* to the edges themselves.

81.　All the projection of the edges may, then, be multiplied by a certain constant, and then will result a drawing on which the lengths of the edges of the cube are given exactly.

82.　This drawing is called the **" Isometric drawing "** of the cube, as distinguished from the Isometric projection.

83.　**To make** an isometric drawing of a cube, the true lengths of the edges are simply laid off in their proper directions on the drawing.

84.　These directions are known from the consideration that the projections of the edges must form a *regular hexagon* with its diagonals.

85.　All lines, then, are either *vertical* or inclined to the horizontal at angles of 30° either way.

86.　Now, it is evident that not only cubes, but **all rectangular objects** as well, may be represented by isometric drawings.

87.　For, any rectangular object may be placed with its edges parallel to those of the cube. Then they will have the same directions on the drawing as those of the cube, and will be equally " foreshortened," just as are the edges of a cube.

88.　These edges may be considered parallel in space to three *rectangular axes*. These axes will have fixed directions on all isometric drawings,—one vertical, one inclined 30° one way, and the other 30° the other way.

89.　They are called the **isometric axes.**

90. All lines of an object parallel to these axés, are known as **Isometric Lines.**

91. All dimensions parallel to isometric lines, are measured off their true length on the drawing.

92. Measurements not parallel to isometric lines, *cannot* be laid off in their true values on the isometric drawing.

93. **Isometric drawing is best adapted** to representing objects made up of plane surfaces, and whose principal lines are parallel to the three rectangular axes.

94. In almost all other cases there is great distortion of the picture.

95. It is, however, sometimes necessary to represent *non-isometric* lines, and to draw curves isometrically, and to represent special points on an isometric drawing.

USE OF CO-ORDINATES.

96. In all such cases it is necessary to have first the orthographic projections of the object. From these the *three rectangular co-ordinates* of any point of the object may be found, referred to some point that can be represented easily, and these then laid off on the drawing.

97. That is, the location of *any point* may be determined by measurements along isometric lines from some point that can be shown isometrically.

EXAMPLE.

98. The use of co-ordinates in general is illustrated by an example in Fig. 6, which is a simple timber joint. Evidently, the horizontal timber can be easily drawn in isometric, but the lines of the inclined piece and the joint are non-isometric.

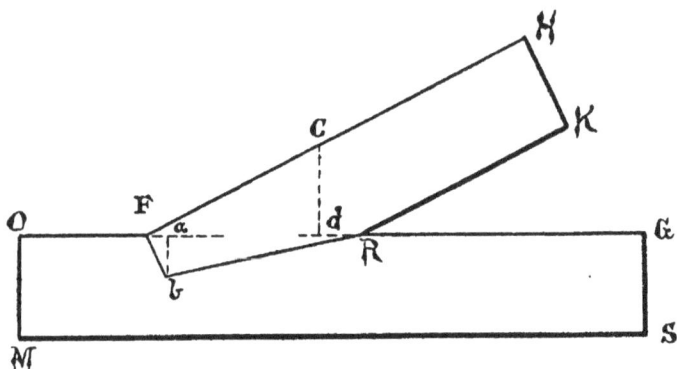

Figure 6.

99.   A single projection is sufficient here.   From it.
Fig. 6, get the " co-ordinates " *Oa* and *ab* of point *b;*
and *Od* and *dc* of any point *C* of the edge *FH*.

Construct the isometric drawing of the horizontal
piece *OS*.   Then along *OG* on the drawing, lay off
Oa from *O*, and from *a* downward lay off *ab*.   This
gives the lines *Fb* and *bR*, as *F* and *R* are on the isometric line and can be laid off at once.

Next, to get the inclination of the other piece, lay off
Od from *O* along *OG*, and upward lay off *dC*.   Then
there are two points determined and *FH* can be drawn.

RK is parallel to *FH* on the object, hence will be on
the drawing.

The widths of the pieces are shown as usual.   In
this way the complete isometric drawing is. readily
constructed.

100.   By a similar use of co-ordinates obtained from
the ordinary projections, the isometric drawing of anything may be constructed.

SHADE LINES.

101.   Shade lines are placed in isometric drawings

according to the same principles as in ordinary projection.

DIRECTION OF LIGHT.

102. The conventional direction of light is as before, from over the left shoulder, in the direction of the body diagonal of a cube. But here, this diagonal is inclined downward and to the right at an angle of 30° with the horizontal.

103. Hence the **direction of light** on an isometric drawing, is 30° downward to the right.

PLATE XII.

Represent by isometric drawing,

(1) A cube with edges $1\frac{1}{4}$ ".

(2) An **H** shaped object formed from a cube $1\frac{1}{4}$ " square, by cutting channels into two opposite faces of the cube, and square in section.

Directions: Arrange in the middle of the plate. Place the shade lines properly.

PLATE XIII.

Construct an isometric drawing of the model of a timber joint, shown in Fig 6, using these dimensions :

Let $OG = 5$ ".

$OM = \frac{7}{8}$ ".

Width, $OW = 1$ ".

$OF = 1\frac{1}{4}$ ".

$FH = 3\frac{1}{4}$ ".

$HK = OM = \frac{3}{4}$ ".

$\dfrac{Cd}{Fd} = \frac{2}{3}$ "

$bFH =$ A right angle.

$ab = \frac{3}{8}$ ".

Directions: Place the drawing to suit the plate. Show the proper shade lines.

OBLIQUE DRAWING.

104. Oblique Drawing is very similar to Isometric. The only difference is that in oblique drawing one of the axes, which in isometric is inclined at 30°, is made *horizontal*, and all lines parallel to it.

105. The lines that are inclined, may make any angle with the horizontal, as well as 30°, and may incline to the left or to the right. 30° and 45° are commonly used, preferably 45°.

SHADE LINES.

106. Shade lines are determined as in Isometric Drawing. When the inclined lines are to the right, the shaded lines are situated on an object just as on the isometric drawing of it. But when the lines incline to the left, they will be different, for the light must come in such a way as to make the *front* face illuminated ; that is, the light cannot come from the rear, hence all the illuminated faces of a cube or prism, or other rectangular object, will be visible.

CABINET PERSPECTIVE.

107. **Cabinet Perspective** is a common term for Oblique Drawing. In this, the inclined lines are drawn at an angle of 45° to the right.

PLATE XIV.

Represent in Oblique Drawing,

(1) A cube 1¼ ″ square, making the inclined lines on the drawing 30° to the right.

(2) The same cube, making the inclined lines 60° the right.

Directions: Place the shade lines properly.

PLATE XV.

Represent in Oblique Drawing,

(1) A cube 1¼ " square, making the inclined lines 45° to the left.

(2) A cube 1¼ " square, making the inclined lines 45° to the right.

Directions: Place the shade lines properly.

Compare the four drawings of this cube in Plates XIV and XV, and decide which represents the cube best.

EXERCISE.

Represent in Cabinet Perspective the timber joint of Plate XIII, using the dimensions there given.

REVIEW QUESTIONS.

Q. 87. What is Isometric Drawing ?

Q. 88. How does an isometric drawing differ from a perspective view ?

Q. 89. How is the isometric projection of a cube obtained ? Illustrate.

Q 90. Why is it that the projections of the edges are all equal ?

Q. 91. What kind of a figure is then formed by them ?

Q. 92. How is the isometric drawing derived from the isometric projection ?

Q. 93. Why is it allowable to multiply all the edges by the same thing ?

Q. 94. How is an isometric drawing of a cube actually made ?

Q. 95. What directions have the edges on the drawing ?

Q. 96. How is it known ?

Q. 97. Can isometric drawings be made of anything but cubes ?

Q. 98. Why ? How ?

Q. 99. What are the isometric axes ?

Q. 100. What are the isometric lines ?

Q. 101. What dimensions may be measured off on an isometric drawing ?

Q. 102. Why cannot others ?

Q. 103. For what are isometric drawings best adapted ?

Q. 104. How are isometric drawings of curves and non-isometric lines in general, drawn ? Explain and illustrate.

Q. 105. How are shade lines determined in isometric drawing ?

Q. 106. What is the conventional direction of light ?

Q. 107. What is the direction of a ray on the drawing ?

Q. 108. How does Oblique Drawing differ from Isometric.

Q. 109. What are the usual directions for the axes in Oblique Drawing ?

Q. 110. How are shade lines determined in Oblique drawing ?

Q. 111. Do the shade lines of a cube come on the same edges when it is shown with the inclined lines to the left, as when they are to the right ?

Q. 112. Why is this so ?

Q. 113. What is Cabinet Perspective ?

Q 114. Why so called ?

Ans. Because it is an approximation to true perspective, and was intended to be especially useful to cabinet makers.

Q. 115. What are the directions of the principal lines in Cabinet Perspective ?

WORKING DRAWINGS.

108. **The object of a working drawing** is, briefly, to show the workman *what to make and how to make it.*

109. Working drawings are made in accordance with the principles of projection, but the rigid laws, and the refinements and conventionalities of Projection Drawing, are not strictly observed. They are made in the simplest and best way to accomplish the object for which they were intended.

THE THREE VIEWS.

110. Usually the ordinary three projections are employed, situated in their customary relation to each other.

111. The vertical projection is called simply **the elevation**, or the "front" or "side view."

112. The horizontal projection is called **the Plan**, or "top view."

113. The third projection is called the **second elevation,** or "end elevation," or simply "end view,"—in some cases, "side view."

114. In some cases **two end views** are needed. The view from the left is drawn to the right of the elevation, and the view from the right to the left of it.

115. When the views are not placed in their usual relation, they must each be plainly marked.

SECTIONS.

116. In very many cases a better idea can be given the workman of the shape and dimensions of an object, if it is cut thro in some way by a plane, and the cut surface drawn. Such drawings are called "sectional views," or "sections."

117. **Sections are used** to show interiors of hollow objects, and in all cases where the clearness of the drawing will be increased.

118. **Sectional views are arranged** in convenient relation to the other views. The best way of taking sections and arranging the views, varies with each particular case, and can best be learned by practice.

119. In many cases, pieces are better sectioned half way thro, and the other half left as a simple view.

120. In very many other cases, it is sufficient to pass a cutting plane only a little way, and then break out a piece to disclose some special feature.

121. As a rule, **cutting planes** are passed parallel to the planes of projection whenever possible.

CROSS SECTIONING.

122. In every sectional view, *wherever material is cut*, the cut surface is "*sectioned*" with fine lines spaced equal distances apart, inclined 45° to the horizontal.

123. **The spacing** will vary with the size of the drawing, being finer for small drawings than for large ones.

124. Cut pieces adjoining on the drawing, should be sectioned in opposite ways. When it is impossible to avoid sectioning two adjacent pieces in the same direction, the ruling should offset. And the smaller piece may be sectioned finer than the other. Thus there is never any cause for confusion.

Note: Some favor the use of 30° and 60° lines for sectioning, when there are several cut surfaces adjacent. This is not necessary, however, and is not so good.

125. Formerly there was a particular kind of cross sectioning employed for every different kind of material.

It is usual now, and much better, to mark each piece by means of reference letters, or in words, to show the kind of material to be used. This can be done for pieces not sectioned, as well as for those that are. There is no chance then for mistakes due to ignorance of the conventionalities on the part of the workmen.

GENERAL RULES.

126. No set of fixed rules can be laid down for making working drawings, but a few general directions should be observed.

127. The views to be shown, of an object, the manner of arranging them, and the character and finish of the drawing, are all largely matters of judgment on the part of the draftsman in each instance. He must be guided by the purpose for which the drawing is intended, and by the best practice.

128. On working drawings, "ground line" is generally omitted. It is well to use it in drawings of small pieces.

129. The "projecting lines," also, are never used except on drawings of small pieces, and in cases where the use of a few of them will add to the clearness of the drawing.

130. **Drawings are lettered** only when necessary for reference, and then simply in whatever way may seem best in the particular case.

SHADE LINES.

131. Shade lines should always be placed on working drawings according to the rules of Projection Drawing. Shade lines *correctly* placed, add greatly, not only to the appearance of a drawing, but also to the ease with which it may be used.

132. Shade lines are determined for sectiónal views in the same way as for other views.

DRAWING TO SCALE.

133. All drawings should be "made to scale," so that the dimensions on the drawing will bear a fixed ratio to those on the object represented.

134. The best scale to adopt, is a matter of judgment in each particular case, and depends on the size of the drawing desired, and the peculiarities of the thing represented. The scale should be chosen so that the drawing will not be unnecessarily large, and yet large enough to show the details of the thing represented, clearly. The scale adopted should be noted on the drawing, as, $1\frac{1}{2}'' = 1'$.

DIMENSIONING DRAWINGS.

135. Altho drawings are made to scale, all important dimensions must be clearly marked on the drawing in figures. It is very important that the figures be perfectly distinct. Feet are indicated by the prime mark ('), and inches by the seconds (").

136. Dimensions that are horizontal, or nearly so, read, of course, from left to right. Dimensions that are vertical, must read from bottom to top.

137. The exact point, or lines, between which the figures give the measurements, are indicated by a line called the "dimension line," drawn between them.

138. The following is used as **the dimension line,** —·—·—·—·—, a neat "arrow head" being placed at each end, the line being broken out for the figures. The length of the dashes used, varies somewhat with the size of the drawing and the use to be made of it.

139. The placing of the dimensions on a drawing, is a matter requiring considerable attention. Every

necessary measurement must be given, but given only once.

The "over all" dimension should always be given, and in close relation to those of the parts.

Care must be taken not to confuse a drawing with the figures or the dimension lines.

CENTER LINE.

140. It is often necessary to use a "center line." Its uses will be learned by practice, as necessity for it occurs. It would be drawn, for instance, in the drawing of a shaft and pulleys, thro the center of the shaft. In the plan of an engine, it would be drawn thro the center of the shaft and also thro the center of the cylinder, perpendicular to the shaft.

141. The line used is like the dimension line, but made up of a dash and *two* dots repeated.

142. The center line is also used to indicate the location of a section. It is drawn where the cutting plane is imagined to be passed.

ISOMETRIC AND OBLIQUE FOR WORKING DRAWINGS.

143. **Isometric and oblique drawings** are not customarily used for working drawings. They may be used, however, for simple things. They are well suited for working drawings of simple rectangular objects, but are not generally used. Their principal use is for purposes of illustration. They serve to make doubly clear anything that may be difficult to understand from the ordinary working drawing. They are helpful to those who are not very ready in reading regular working drawings. In such cases, they ac-

company the regular working drawings, and are not dimensioned.

GENERAL VIEWS AND DETAIL DRAWINGS.

144. A drawing showing all the parts of a machine or other structure, put together in their proper places, is called the "general drawing," or "general view."

145. When a machine is being designed, general views are made, usually plan, elevation, and second elevation, and sometimes sections. These show all the pieces in their proper relations to each other, drawn to scale, so that the designer is able to adapt them to each other, and decide upon their dimensions.

DETAILS.

146. These general drawings are usually too much crowded and confused to allow the dimensions of all the pieces to be placed on them, so that the mechanic can work from them.

147. Each piece must be picked out from the general drawing, and a separate and larger drawing made of it to scale, with the dimensions all figured. These drawings of the pieces are called "Detail Drawings," or "Details."

148. In cases where much machine work is to be done on the pieces, a set of details is made for the pattern maker, and another set, showing the dimensions of the finished pieces and the machine work required, is made for the machinist. The instructions to both the pattern maker and the machinist, may often be shown on one drawing.

149. **Detail drawings are made** in pencil on ordinary paper, preferably "detail paper," and then tracings made, which may be preserved. Prints are made for the workmen.

INDEXING.

150. Details of complicated machines must be plainly marked, and numbered and indexed according to some system, so that the tracing of any detail may be readily found.

151. Further knowledge of the endless variety of methods followed by draughtsmen in making working drawings can only be acquired by prolonged practice under the direction of some one competent to instruct. In time, the "*idea*" of it all will become so instilled into one that he can work out ways for himself to fit any case he may be called upon to handle.

REVIEW QUESTIONS.

Q. 116. What is the object of working drawings ?
Q. 117. How are working drawings made ?
Q. 118. Are any rigid set of rules followed strictly ?
Q. 119. What is to guide then, in making a working drawing for a particular purpose ?
Q. 120. What is the common name for the vertical projection ?
Q. 121. For the horizontal projection ?
Q. 122. When two end views are needed, how are they placed ?
Q. 123. Is it allowable to place the views in any other relation ?
Ans. Yes, if desirable for any special reason. Each view may be on a separate sheet, if necessary. (See Sec. 115.)
Q. 124. What are sections ?
Q. 125. For what are they used ?
Q. 126. How should the cutting plane be passed ?
Q. 127. Can any fixed rules be given for taking sections ?
Q. 128. What is to guide, then ?
Q. 129. How are cut surfaces shown on a drawing ?
Q. 130. How is cross sectioning done ?
Q. 131. Are there any fixed rules that can be laid down for making working drawings, to cover all cases ?
Q. 132. Why ?
Q. 133. What are some general rules that should be observed ?
Q. 134. Are shade lines used on working drawings ?
Q. 135. How are the shade lines determined for sections ?
Q. 136. What governs the choice of the best scale for a drawing ?
Q. 137. What are some rules to be observed in dimensioning drawings ?

Q 138. What is the conventional dimension line ?

Q. 139. For what purposes is a center line used ?

Q. 140. What is the conventional center line ?

Q. 141. Are isometric and oblique drawings used as working drawings ?

Q. 142. Are they suitable for working drawings ?

Q. 143. How are they generally used, when at all ?

Q. 144. What is meant by a general view ?

Q 145. When is such a drawing required ?

Q. 146. What are detail drawings ?

Q. 147. How are detail drawings usually prepared ?

Q. 148. How can it be arranged so that the shop drawing of any particular detail of a large complicated machine may be readily found ?

Q. 149. What may be said of acquiring familiarity with the many little " ways " of practical draughtsmen ?

ERRATA.

1. Page 22, Section 43 : The reference should be to page 19, instead of 13.

2. Page 23 : Sub-section (4) should read as follows :
" (4) If a line is perpendicular to either co-ordinate plane, its projection on that plane is a *point*, and its projection on the other plane is a straight line perpendicular to the ground line."

3. Page 24, sub-section (7) : For "two planes in space," read "two *lines* in space."

4. Page 25, section 49 : For " whose center is the axis," read " whose center is *in* the axis."

5. Page 29, *Directions* : For " whenever," read " wherever."

6. Page 33, Q. 48 : For " project," read " projection."

7. Page 33, Q. 60 : For " moves," read " revolves."

8. Page 34, Q. 77 : For " Complete," read " completely."

9. Page 35, section 78 : The last sentence should read, " Hence their V projections are all equal."

10. Page 36, section 81 : In the first line, for " projection," read " projections." In the second line, for " then," read " there."

11. Page 38, section 99 : For *dc* read *d'C*.

12. Page 39 : Problem (2), under Plate XII, should read as follows :
"(2) An H–shaped object formed from a cube 1¼ " square, by cutting channels into two opposite faces of the cube, the channels to be one-third of the width of the cube, and square in section."

13. Page 40, section 106 : For "shaded lines," read "shade lines."

14. Page 41, Q. 100 : Omit "the."

15. Page 45, section 128 : Insert "the " before "ground line."

16. Page 45 : The last word on the page should be " read," instead of " used."